The Art of War

The Art of War

Sun Tzu

This edition published in 2019 by Arcturus Publishing Limited
26/27 Bickels Yard, 151–153 Bermondsey Street,
London SE1 3HA

AD005628UK

Printed in the UK

Contents

INTRODUCTION

Sun Tzu's *The Art of War* is probably the earliest-known treatise on war and military science. It is certainly the most influential. Essential reading for strategists in the East since ancient times, it is thought that Napoleon read *The Art of War* when a new edition was published in Paris in 1782.

It was also picked up by two great military theorists who studied Bonaparte's methods – the Prussian Carl von Clausewitz, whose *On War* was published posthumously in 1832, and the French General Baron Antoine-Henri de Jomini, whose *Summary of the Art of War* appeared in 1838.

Also influenced were the pre-Second World War theorists of tank warfare General John Fuller and B.H. Liddell Hart – and, consequently, the German Panzer leaders, including Heinz Guderian, who devoured their work. Field Marshal Bernard Montgomery, victor of El Alamein and commander of land forces on D-Day, wrote extensively

on *The Art of War* in his monumental *A History of Warfare*. His great rival, General George Patton, is thought to have read Sun Tzu as a teenager in the Virginia Military Institute before he went to West Point.

General Douglas MacArthur and Joseph Stalin were also said to have been fans of Sun Tzu. *The Art of War* was adopted as a handbook on guerrilla warfare by Mao Zedong, Fidel Castro and Vo Nguyen Giap, the Vietnamese strategist who beat the French and the Americans in Indo-China, and according to US Marine Corps General Paul K. Van Riper, it influenced the planning of Operation Desert Storm.

The first mention of Sun Tzu as a great military strategist was made, in passing, by the Taoist philosopher Huai-nan Tzu who died in 122BC. The 'Tzu' here, as in the case of Sun Tzu, is an honorific that means 'master'. Sun Tzu – Master Sun – is known more commonly in the literature by his personal name Sun Wu, though 'Wu' might be a nickname as it means 'military'. The surname 'Sun' was said to have been bestowed on Sun Tzu's grandfather by Duke Ching of Ch`i. Ch`i was Sun Tzu's home state which he fled after a rebellion. Sun Tzu would have called himself Ch`ang-ch`ing. This was a so-called 'style', a new name Chinese males of that era adopted at the age of 20.

Sun Tzu fled from Ch`i to the nearby state of Wu where, the great Han historian Ssu-ma Ch'ien (145–85BC) says, *The Art of War* brought him to the attention of Ho Lu, the king of Wu. This must have been before 512BC, as Ssu-ma Ch'ien also records that Sun Tzu was with Ho Lu when Wu attacked the kingdom of Ch'u that year. Ssu-ma Ch'ien also related the story of how Sun Tzu convinced the king of his military prowess.

'I have carefully perused your 13 chapters,' said Ho Lu, king of Wu. 'May I submit your theory of managing soldiers to a slight test?'

Sun Tzu replied: 'You may.'

Ho Lu asked: 'May the test be applied to women?'

The answer again was 'Yes', so arrangements were made to bring 180 ladies out of the palace.

Sun Tzu divided them into two companies, placing one of the king's favourite concubines at the head of each. He then asked them all to take spears in their hands and said: 'I presume you know the difference between front and back, right hand and left hand?'

The women replied: 'Yes.'

'When I say: "Eyes front," you must look straight ahead,' Sun Tzu continued. 'When I say: "Left turn," you must face towards your left hand. When I say: "Right turn," you must face towards your right hand. When I say: "About turn," you must face right round towards your back.'

Having explained the words of command, Sun Tzu began the drill. Then, to the sound of drums, he gave the order, 'Right turn'. But the girls only burst out laughing.

Sun Tzu said: 'If words of command are not clear and distinct, if orders are not thoroughly understood, then the general is to blame.'

So he started drilling them again. This time he gave

the order, 'Left turn'. Again the girls burst into fits of laughter. Sun Tzu said again: 'If words of command are not clear and distinct, if orders are not thoroughly understood, the general is to blame. But if his orders are clear, and the soldiers disobey, then it is the fault of their officers.'

So he ordered the leaders of the two companies to be beheaded.

The king of Wu was watching from the top of a raised pavilion. When he saw that his favourite concubines were about to be executed, he quickly sent a message, saying: 'We are now quite satisfied as to our general's ability to handle troops. If we are bereft of these two concubines, our meat and drink will lose their savour. It is our wish that they shall not be beheaded.'

But Sun Tzu replied: 'Having once received His Majesty's commission to be the general of his forces, there are certain commands of His Majesty which, acting in that capacity, I am unable to accept.'

Accordingly, he had the two concubines beheaded and installed another pair as leaders in their place. Then the drum was sounded for the drill again. This time the girls turned right and left, marching ahead or wheeling back, knelt or stood as ordered, all with perfect accuracy and precision, and without venturing a sound.

Then Sun Tzu sent a messenger to the king, saying: 'Your soldiers, Sire, are now properly drilled and disciplined, and ready for Your Majesty's inspection. They

> can be put to any use that their sovereign may desire; bid them go through fire and water, and they will not disobey...'
>
> After that, Ho Lu saw that Sun Tzu knew how to handle an army, and finally made him a general. In the west, he defeated the Ch`u State and forced his way into Ying, the capital; to the north he put fear into the States of Ch`i and Chin, and spread his fame abroad amongst the feudal princes. And Sun Tzu shared in the might of the king.

In the introduction to his famous 1910 translation of *The Art of War* reprinted here, Lionel Giles cast doubt on this account. He pointed out that the great contemporary chronicle of this period, the Tso Chuan, failed to mention Sun Tzu among the Wu generals that invaded Ch`u, though lesser figures are mentioned. This has led some to conclude that Sun Tzu did not exist and that *The Art of War* is a compilation of other authors.

From his intimate knowledge of the text, Giles concluded that *The Art of War* was the work of a single author and pointed out that, as a foreigner in Wu, Sun Tzu would not have had the civil rank that was needed to become a general at that time. Clearly, from his writing, the author had practical experience of warfare and had been involved in the attack on Ch`u in 512BC and the capture of Ying in 506BC. However, in the text he mentions conflict with the Yueh. The Wu first attacked the Yueh in 510BC, so *The Art of War* could not have been written before 512BC as Ssu-ma Ch`ien maintained. The Yueh struck back when the Wu were in Ch`u in 505BC. The Wu then counterattacked in 496BC, but were defeated and Ho Lu was killed.

In Chapter VI of *The Art of War*, Sun Tzu writes: 'Though according to my estimate the soldiers of Yueh exceed our own in number, that shall advantage them nothing in the matter of victory. I say then that victory can be achieved.'

This sentence, Giles maintained, was hardly 'one that could have been written in the full flush of victory'. As Ho Lu died in 496BC, if the book was written for him, it must have been written during the period 505–496BC, when there was a lull in the hostilities. Sun Tzu's name does not appear in the historical record again after the death of Ho Lu, but there is no reason to believe that he did not survive to participate in the short-lived resurgence of the Wu who, under Fu Ch`ai, captured the capital of the Yueh in 494BC. The Yueh hit back in 482BC, finally destroying the state of Wu in 473BC. Giles believed that *The Art of War* could also have been written between 496 and 494BC, or 482 and 473BC Ssu-ma Ch`ien's sources might be confused and the drilling of the concubines could have been a preliminary exercise for the work. It is also possible that Ho Lu saw an early draft of the manuscript before 512BC, which was revised later after hostilities with the Yueh broke out.

While Ssu-ma Ch`ien mentioned '13 chapters' and 13 chapters are presented here, some early accounts of *The Art of War* said that there were 82 *p`ien* or chapters.

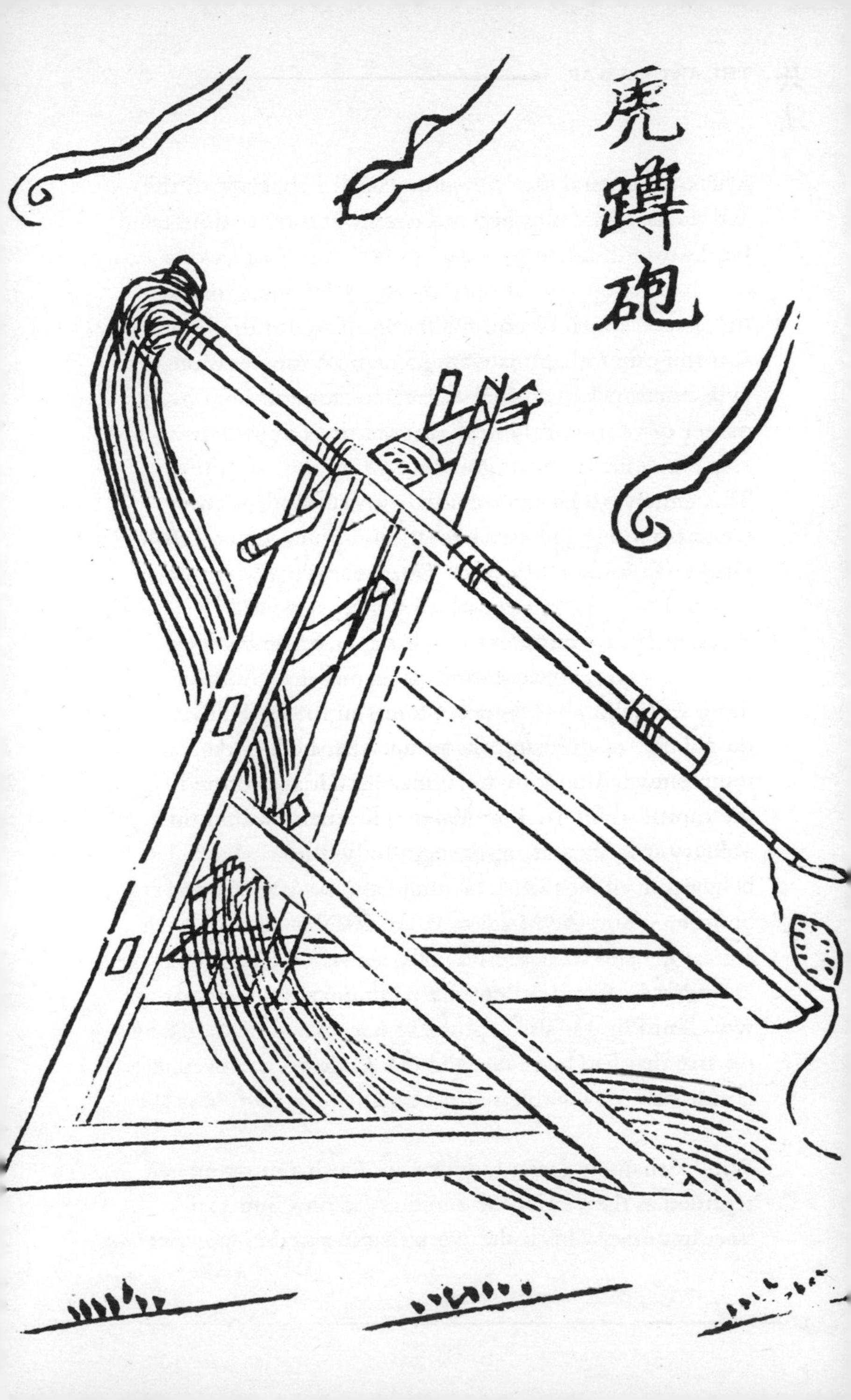
虎蹲砲

There are several theories about what happened to the other chapters. Giles pointed out that in some shorter works *p`ien* is taken to mean 'leaves'. Sun Tzu's work may have been bound with the work of other military theorists, or various commentaries. It is also thought that the other chapters were records of conversations between Sun Tzu and Ho Lu, discussing the finer points of strategy. One of these appears in the T'ung-tien, an encyclopedic work on government compiled in the 8th century AD by the mandarin Tu Yu. Called 'Nine Configurations and Two Questions', it appears as a 14th chapter in some editions of *The Art of War*.

Then in 1972 a number of texts engraved on bamboo were found in a grave in Shandong province. Among them were the 13 chapters of Sun Tzu's *Art of War*, though the manuscript was a millennium older than any then known. Alongside were another 33 chapters, but these concerned the practicalities of warfare rather than military strategy and are thought to be the work of a descendant of Sun Tzu's, possibly his grandson, named Sun Pin – 'Sun the Mutilated'. He was said to have been a great general who, according to Ssu-ma Ch`ien, 'had his feet cut off and yet continued to discuss the art of war'. Sun Pin was also reputed to have written a book on warfare that had been lost and the 33 newly discovered chapters are now published as Sun Pin's *The Lost Art of War*.

The 13 chapters presented here are those that have been regarded as the work of the military genius Sun Tzu since antiquity. This is the work that inspired Napoleon,

Mao Zedong *et al.* And it is this classic 1910 translation that would have been pored over by Fuller, Liddell Hart, Montgomery and Patton, while the Allies faced Panzer commanders who had drawn their inspiration from the same source.

In the West, *The Art of War* now has a readership outside military strategists. One of Sun Tzu's great concerns was the politics of war, so it is essential reading for politicians, diplomats and those involved in international relations. Business gurus now teach *The Art of War*, using Sun Tzu's view of warfare as a metaphor for the struggle for global markets. To the Chinese, though, it remains one of the great pillars of classical literature.

For some 2,500 years, Sun Tzu's treatise has had something profound to say about the human condition. The world is no less warlike than it was when the Chinese were slugging it out in the Spring and Autumn (770–476BC) and the Warring States (475–221BC) periods, before the first emperor Shih huang-ti united China, so undoubtedly Sun Tzu has something to say to us now.

It is also ironic to note that Sun Tzu, perhaps the greatest writer on war, was alive and at work at the same time as China's most illustrious man of peace, the great sage Confucius.

NIGEL CAWTHORNE

CHAPTER ONE

LAYING PLANS

1 Sun Tzu said: The art of war is of vital importance to the State.

2 It is a matter of life and death, a road either to safety or to ruin. Hence it is a subject of inquiry which can on no account be neglected.

3 The art of war, then, is governed by five constant factors, to be taken into account in one's deliberations, when seeking to determine the conditions obtaining in the field.

4 These are:
(1) The Moral Law;
(2) Heaven;
(3) Earth;
(4) The Commander;
(5) Method and Discipline.

5 & 6 The Moral Law causes the people to be in complete accord with their ruler, so that they will follow him regardless of their lives, undismayed by any danger.

7 Heaven signifies night and day, cold and heat, times and seasons.

8 Earth comprises distances, great and small; danger and security; open ground and narrow passes; the chances of life and death.

9 The Commander stands for the virtues of wisdom, sincerity, benevolence, courage and strictness.

10 By Method and Discipline are to be understood the marshalling of the army in its proper subdivisions, the graduations of rank among the officers, the maintenance of roads by which supplies may reach the army, and the control of military expenditure.

11 These five heads should be familiar to every general: he who knows them will be victorious; he who knows them not will fail.

12 Therefore, in your deliberations, when seeking to determine the military conditions, let them be made the basis of a comparison, in this wise:—

13 (1) Which of the two sovereigns is imbued with the Moral Law?
(2) Which of the two generals has most ability?
(3) With whom lie the advantages derived from Heaven and Earth?
(4) On which side is Discipline most rigorously enforced?
(5) Which army is stronger?
(6) On which side are officers and men more highly trained?
(7) In which army is there the greater constancy both in reward and punishment?

14 By means of these seven considerations I can forecast victory or defeat.

15 The general that hearkens to my counsel and acts upon it, will conquer: let such a one be retained in command! The general that hearkens not to my counsel nor acts upon it, will suffer defeat:— let such a one be dismissed!

16 While heeding the profit of my counsel, avail yourself also of any helpful circumstances over and beyond the ordinary rules.

17 According as circumstances are favourable, one should modify one's plans.

18 All warfare is based on deception.

19 Hence, when able to attack, we must seem unable; when using our forces, we must seem inactive; when we are near, we must make the enemy believe we are far away; when far away, we must make him believe we are near.

20 Hold out baits to entice the enemy. Feign disorder, and crush him.

21 If he is secure at all points, be prepared for him. If he is in superior strength, evade him.

22 If your opponent is of choleric temper, seek to irritate him. Pretend to be weak, that he may grow arrogant.

23 If he is taking his ease, give him no rest. If his forces are united, separate them.

24 Attack him where he is unprepared, appear where you are not expected.

25 These military devices, leading to victory, must not be divulged beforehand.

26 Now the general who wins a battle makes many calculations in his temple ere the battle is fought. The general who loses a battle makes but few calculations beforehand. Thus do many calculations lead to victory, and few calculations to defeat: how much more no calculation at all! It is by attention to this point that I can foresee who is likely to win or lose.

CHAPTER TWO

WAGING WAR

1 Sun Tzu said: In the operations of war, where there are in the field a thousand swift chariots, as many heavy chariots, and a hundred thousand mail-clad soldiers, with provisions enough to carry them a thousand *li**, the expenditure at home and at the front, including entertainment of guests, small items such as glue and paint, and sums spent on chariots and armour, will reach the total of a thousand ounces of silver per day. Such is the cost of raising an army of 100,000 men.

2 When you engage in actual fighting, if victory is long in coming, then men's weapons will grow dull and their ardour will be damped. If you lay siege to a town, you will exhaust your strength.

3 Again, if the campaign is protracted, the resources of the State will not be equal to the strain.

4 Now, when your weapons are dulled, your ardour damped, your strength exhausted and your treasure spent, other chieftains will spring up to take advantage of your extremity. Then no man, however wise, will be able to avert the consequences that must ensue.

* One *li* is equal to half a kilometre.

5 Thus, though we have heard of stupid haste in war, cleverness has never been seen associated with long delays.

6 There is no instance of a country having benefited from prolonged warfare.

7 It is only one who is thoroughly acquainted with the evils of war that can thoroughly understand the profitable way of carrying it on.

8 The skilful soldier does not raise a second levy, neither are his supply-wagons loaded more than twice.

9 Bring war material with you from home, but forage on the enemy. Thus the army will have food enough for its needs.

10 Poverty of the State Exchequer causes an army to be maintained by contributions from a distance. Contributing to maintain an army at a distance causes the people to be impoverished.

11 On the other hand, the proximity of an army causes prices to go up; and high prices cause the people's substance to be drained away.

12 When their substance is drained away, the peasantry will be afflicted by heavy exactions.

13 & 14 With this loss of substance and exhaustion of strength, the homes of the people will be stripped bare, and three-tenths of their income will be dissipated; while government expenses for broken chariots, worn-out horses, breast-plates and helmets, bows and arrows, spears and shields, protective mantles, draught-oxen and heavy wagons, will amount to four-tenths of its total revenue.

15 Hence a wise general makes a point of foraging on the enemy. One cartload of the enemy's provisions is equivalent to twenty of one's own, and likewise a single *picul** of his provender is equivalent to twenty from one's own store.

16 Now in order to kill the enemy, our men must be roused to anger; that there may be advantage from defeating the enemy, they must have their rewards.

* One *picul* weighs approximately 133 lbs.

17 Therefore in chariot fighting, when ten or more chariots have been taken, those should be rewarded who took the first. Our own flags should be substituted for those of the enemy, and the chariots mingled and used in conjunction with ours. The captured soldiers should be kindly treated and kept.

18 This is called using the conquered foe to augment one's own strength.

19 In war, then, let your great object be victory, not lengthy campaigns.

20 Thus it may be known that the leader of armies is the arbiter of the people's fate, the man on whom it depends whether the nation shall be in peace or in peril.

CHAPTER THREE

ATTACK STRATAGEMS

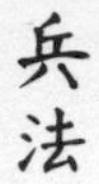

1 Sun Tzu said: In the practical art of war, the best thing of all is to take the enemy's country whole and intact; to shatter and destroy it is not so good. So, too, it is better to recapture an army entire than to destroy it, to capture a regiment, a detachment or a company entire than to destroy them.

2 Hence to fight and conquer in all your battles is not supreme excellence; supreme excellence consists in breaking the enemy's resistance without fighting.

3 Thus the highest form of generalship is to balk the enemy's plans; the next best is to prevent the junction of the enemy's forces; the next in order is to attack the enemy's army in the field; and the worst policy of all is to besiege walled cities.

4 The rule is, not to besiege walled cities if it can possibly be avoided. The preparation of mantlets, movable shelters, and various implements of war, will take up three whole months; and the piling up of mounds against the walls will take three months more.

5 The general, unable to control his irritation, will launch his men to the assault like swarming ants, with the result that one-third of his men are slain, while the town still remains untaken. Such are the disastrous effects of a siege.

6 Therefore the skilful leader subdues the enemy's troops without any fighting; he captures their cities without laying siege to them; he overthrows their kingdom without lengthy operations in the field.

7 With his forces intact he will dispute the mastery of the Empire, and thus, without losing a man, his triumph will be complete. This is the method of attacking by stratagem.

8 It is the rule in war, if our forces are ten to the enemy's one, to surround him; if five to one, to attack him; if twice as numerous, to divide our army into two.

9 If equally matched, we can offer battle; if slightly inferior in numbers, we can avoid the enemy; if quite unequal in every way, we can flee from him.

10 Hence, though an obstinate fight may be made by a small force, in the end it must be captured by the larger force.

11 Now the general is the bulwark of the State; if the bulwark is complete at all points, the State will be strong; if the bulwark is defective, the State will be weak.

12 There are three ways in which a ruler can bring misfortune upon his army:—

13 (1) By commanding the army to advance or to retreat, being ignorant of the fact that it cannot obey. This is called hobbling the army.

14 (2) By attempting to govern an army in the same way as he administers a kingdom, being ignorant of the conditions which obtain in an army. This causes restlessness in the soldier's minds.

15 (3) By employing the officers of his army without discrimination, through ignorance of the military principle of adaptation to circumstances. This shakes the confidence of the soldiers.

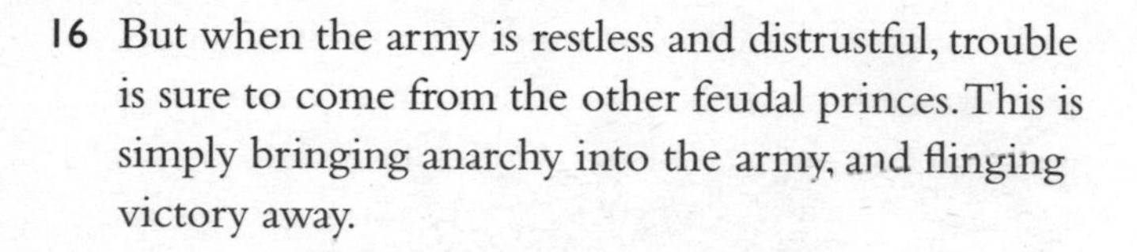

16 But when the army is restless and distrustful, trouble is sure to come from the other feudal princes. This is simply bringing anarchy into the army, and flinging victory away.

17 Thus we may know that there are five essentials for victory:

(1) He will win who knows when to fight and when not to fight.

(2) He will win who knows how to handle both superior and inferior forces.

(3) He will win whose army is animated by the same spirit throughout all its ranks.

(4) He will win who, prepared himself, waits to take the enemy unprepared.

(5) He will win who has military capacity and is not interfered with by the sovereign.

18 Hence the saying: If you know the enemy and know yourself, you need not fear the result of a hundred battles. If you know yourself but not the enemy, for every victory gained you will also suffer a defeat. If you know neither the enemy nor yourself, you will succumb in every battle.

CHAPTER FOUR

TACTICAL DISPOSITIONS

1 Sun Tzu said: The good fighters of old first put themselves beyond the possibility of defeat, and then waited for an opportunity of defeating the enemy.

2 To secure ourselves against defeat lies in our own hands, but the opportunity of defeating the enemy is provided by the enemy himself.

3 Thus the good fighter is able to secure himself against defeat, but cannot make certain of defeating the enemy.

4 Hence the saying: One may know how to conquer without being able to do it.

5 Security against defeat implies defensive tactics; ability to defeat the enemy means taking the offensive.

6 Standing on the defensive indicates insufficient strength; attacking, a superabundance of strength.

7 The general who is skilled in defence hides in the most secret recesses of the earth; he who is skilled in attack flashes forth from the topmost heights of heaven. Thus on the one hand we have the ability to protect ourselves; on the other, a victory that is complete.

8 To see victory only when it is within the ken of the common herd is not the acme of excellence.

9 Neither is it the acme of excellence if you fight and conquer and the whole Empire says: 'Well done!'

10 To lift an autumn leaf is no sign of great strength; to see the sun and moon is no sign of sharp sight; to hear the noise of thunder is no sign of a quick ear.

11 What the ancients called a clever fighter is one who not only wins, but excels in winning with ease.

12 Hence his victories bring him neither reputation for wisdom nor credit for courage.

13 He wins his battles by making no mistakes. Making no mistakes is what establishes the certainty of victory, for it means conquering an enemy that is already defeated.

14 Hence the skilful fighter puts himself into a position which makes defeat impossible, and does not miss the moment for defeating the enemy.

15 Thus it is that in war the victorious strategist only seeks battle after the victory has been won, whereas he who is destined to defeat first fights and afterwards looks for victory.

16 The consummate leader cultivates the Moral Law, and strictly adheres to method and discipline; thus it is in his power to control success.

17 In respect of military method, we have, firstly, Measurement; secondly, Estimation of quantity; thirdly, Calculation; fourthly, Balancing of chances; fifthly, Victory.

18 Measurement owes its existence to Earth; Estimation of quantity to Measurement; Calculation to Estimation of quantity; Balancing of chances to Calculation; and Victory to Balancing of chances.

19 A victorious army opposed to a routed one, is as a pound's weight placed in the scale against a single grain.

20 The onrush of a conquering force is like the bursting of pent-up waters into a chasm a thousand fathoms deep.

CHAPTER FIVE
ENERGY

1 Sun Tzu said: The control of a large force is the same principle as the control of a few men: it is merely a question of dividing up their numbers.

2 Fighting with a large army under your command is nowise different from fighting with a small one: it is merely a question of instituting signs and signals.

3 To ensure that your whole host may withstand the brunt of the enemy's attack and remain unshaken – this is effected by manoeuvres direct and indirect.

4 That the impact of your army may be like a grindstone dashed against an egg – this is effected by the science of weak points and strong.

5 In all fighting, the direct method may be used for joining battle, but indirect methods will be needed in order to secure victory.

6 Indirect tactics, efficiently applied, are inexhaustible as Heaven and Earth, unending as the flow of rivers and streams; like the sun and moon, they end but to begin anew; like the four seasons, they pass away to return once more.

7 There are not more than five musical notes, yet the combinations of these five give rise to more melodies than can ever be heard.

8 There are not more than five primary colours (blue, yellow, red, white and black), yet in combination they produce more hues than can ever be seen.

9 There are not more than five cardinal tastes (sour, acrid, salt, sweet and bitter), yet combinations of them yield more flavours than can ever be tasted.

10 In battle, there are not more than two methods of attack – the direct and the indirect; yet these two in combination give rise to an endless series of manoeuvres.

11 The direct and the indirect lead on to each other in turn. It is like moving in a circle – you never come to an end. Who can exhaust the possibilities of their combination?

12 The onset of troops is like the rush of a torrent which will even roll stones along in its course.

13 The quality of decision is like the well-timed swoop of a falcon which enables it to strike and destroy its victim.

14 Therefore the good fighter will be terrible in his onset, and prompt in his decision.

15 Energy may be likened to the bending of a crossbow; decision, to the releasing of a trigger.

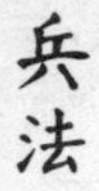

16 Amid the turmoil and tumult of battle, there may be seeming disorder and yet no real disorder at all; amid confusion and chaos, your array may be without head or tail, yet it will be proof against defeat.

17 Simulated disorder postulates perfect discipline, simulated fear postulates courage; simulated weakness postulates strength.

18 Hiding order beneath the cloak of disorder is simply a question of subdivision; concealing courage under a show of timidity presupposes a fund of latent energy; masking strength with weakness is to be effected by tactical dispositions.

19 Thus one who is skilful at keeping the enemy on the move maintains deceitful appearances, according to which the enemy will act. He sacrifices something, that the enemy may snatch at it.

20 By holding out baits, he keeps him on the march; then with a body of picked men he lies in wait for him.

21 The clever combatant looks to the effect of combined energy, and does not require too much from individuals. Hence his ability to pick out the right men and utilize combined energy.

22 When he utilizes combined energy, his fighting men become as it were like unto rolling logs or stones. For

it is the nature of a log or stone to remain motionless on level ground, and to move when on a slope; if four-cornered, to come to a standstill, but if round-shaped, to go rolling down.

23 Thus the energy developed by good fighting men is as the momentum of a round stone rolled down a mountain thousands of feet in height. So much on the subject of energy.

CHAPTER SIX

WEAK POINTS AND STRONG

1 Sun Tzu said: Whoever is first in the field and awaits the coming of the enemy, will be fresh for the fight; whoever is second in the field and has to hasten to battle will arrive exhausted.

2 Therefore the clever combatant imposes his will on the enemy, but does not allow the enemy's will to be imposed on him.

3 By holding out advantages to him, he can cause the enemy to approach of his own accord; or, by inflicting damage, he can make it impossible for the enemy to draw near.

4 If the enemy is taking his ease, he can harass him; if well supplied with food, he can starve him out; if quietly encamped, he can force him to move.

5 Appear at points which the enemy must hasten to defend; march swiftly to places where you are not expected.

6 An army may march great distances without distress, if it marches through country where the enemy is not.

7 You can be sure of succeeding in your attacks if you only attack places which are undefended. You can ensure the safety of your defence if you only hold positions that cannot be attacked.

8 Hence that general is skilful in attack whose opponent does not know what to defend; and he is skilful in defence whose opponent does not know what to attack.

9 O divine art of subtlety and secrecy! Through you we learn to be invisible, through you inaudible; and hence we can hold the enemy's fate in our hands.

10 You may advance and be absolutely irresistible, if you make for the enemy's weak points; you may retire and be safe from pursuit if your movements are more rapid than those of the enemy.

11 If we wish to fight, the enemy can be forced to an engagement even though he be sheltered behind a high rampart and a deep ditch. All we need do is attack some other place that he will be obliged to relieve.

12 If we do not wish to fight, we can prevent the enemy from engaging us even though the lines of our encampment be merely traced out on the ground. All we need do is to throw something odd and unaccountable in his way.

13 By discovering the enemy's dispositions and remaining invisible ourselves, we can keep our forces concentrated, while the enemy's must be divided.

14 We can form a single united body, while the enemy must split up into fractions. Hence there will be a whole pitted against separate parts of a whole, which means that we shall be many to the enemy's few.

15 And if we are able thus to attack an inferior force with a superior one, our opponents will be in dire straits.

16 The spot where we intend to fight must not be made known; for then the enemy will have to prepare against a possible attack at several different points; and his forces being thus distributed in many directions, the numbers we shall have to face at any given point will be proportionately few.

17 For should the enemy strengthen his van, he will weaken his rear; should he strengthen his rear, he will weaken his van; should he strengthen his left, he will weaken his right; should he strengthen his right, he will weaken his left. If he sends reinforcements everywhere, he will everywhere be weak.

18 Numerical weakness comes from having to prepare against possible attacks; numerical strength, from compelling our adversary to make these preparations against us.

19 Knowing the place and the time of the coming battle, we may concentrate from the greatest distances in order to fight.

20 But if neither time nor place be known, then the left wing will be impotent to succour the right, the right equally impotent to succour the left, the van unable to relieve the rear, or the rear to support the van. How much more so if the furthest portions of the army are anything under a hundred *li* apart, and even the nearest are separated by several *li*!

21 Though according to my estimate the soldiers of Yueh exceed our own in number, that shall advantage them nothing in the matter of victory. I say then that victory can be achieved.

22 Though the enemy be stronger in numbers, we may prevent him from fighting. Scheme so as to discover his plans and the likelihood of their success.

23 Rouse him, and learn the principle of his activity or inactivity. Force him to reveal himself, so as to find out his vulnerable spots.

24 Carefully compare the opposing army with your own, so that you may know where strength is superabundant and where it is deficient.

25 In making tactical dispositions, the highest pitch you can attain is to conceal them; conceal your dispositions, and you will be safe from the prying of the subtlest spies, from the machinations of the wisest brains.

26 How victory may be produced for them out of the enemy's own tactics – that is what the multitude cannot comprehend.

27 All men can see the tactics whereby I conquer, but what none can see is the strategy out of which victory is evolved.

28 Do not repeat the tactics which have gained you one victory, but let your methods be regulated by the infinite variety of circumstances.

29 Military tactics are like unto water; for water in its natural course runs away from high places and hastens downwards.

30 So in war, the way is to avoid what is strong and to strike at what is weak.

31 Water shapes its course according to the nature of the ground over which it flows; the soldier works out his victory in relation to the foe whom he is facing.

32 Therefore, just as water retains no constant shape, so in warfare there are no constant conditions.

33 He who can modify his tactics in relation to his opponent and thereby succeed in winning, may be called a heaven-born captain.

34 The five elements (water, fire, wood, metal, earth) are not always equally predominant; the four seasons make way for each other in turn. There are short days and long; the moon has its periods of waning and waxing.

CHAPTER SEVEN

MANOEUVRING

1 Sun Tzu said: In war, the general receives his commands from the sovereign.

2 Having collected an army and concentrated his forces, he must blend and harmonize the different elements thereof before pitching his camp.

3 After that, comes tactical manoeuvring, than which there is nothing more difficult. The difficulty of tactical manoeuvring consists in turning the devious into the direct, and misfortune into gain.

4 Thus, to take a long and circuitous route, after enticing the enemy out of the way, and though starting after him, to contrive to reach the goal before him, shows knowledge of the artifice of deviation.

5 Manoeuvring with an army is advantageous; with an undisciplined multitude, most dangerous.

6 If you set a fully equipped army in march in order to snatch an advantage, the chances are that you will be too late. On the other hand, to detach a flying column for the purpose involves the sacrifice of its baggage and stores.

7 Thus, if you order your men to roll up their buff-coats, and make forced marches without halting day or night, covering double the usual distance at a stretch, doing a hundred *li* in order to wrest an

advantage, the leaders of all your three divisions will fall into the hands of the enemy.

8 The stronger men will be in front, the jaded ones will fall behind, and on this plan only one-tenth of your army will reach its destination.

9 If you march fifty *li* in order to outmanoeuvre the enemy, you will lose the leader of your first division, and only half your force will reach the goal.

10 If you march thirty *li* with the same object, two-thirds of your army will arrive.

11 We may take it then that an army without its baggage-train is lost; without provisions it is lost; without bases of supply it is lost.

12 We cannot enter into alliances until we are acquainted with the designs of our neighbours.

13 We are not fit to lead an army on the march unless we are familiar with the face of the country – its mountains and forests, its pitfalls and precipices, its marshes and swamps.

14 We shall be unable to turn natural advantage to account unless we make use of local guides.

15 In war, practise dissimulation, and you will succeed.

16 Whether to concentrate or to divide your troops must be decided by circumstances.

17 Let your rapidity be that of the wind, your compactness be that of the forest.

18 In raiding and plundering be like fire, in immovability like a mountain.

19 Let your plans be dark and impenetrable as night, and when you move, fall like a thunderbolt.

20 When you plunder a countryside, let the spoil be divided amongst your men; when you capture new territory, cut it up into allotments for the benefit of the soldiery.

21 Ponder and deliberate before you make a move.

22 He will conquer who has learnt the artifice of deviation. Such is the art of manoeuvring.

23 The Book of Army Management says: On the field of battle, the spoken word does not carry far enough: hence the institution of gongs and drums. Nor can ordinary objects be seen clearly enough: hence the institution of banners and flags.

24 Gongs and drums, banners and flags, are means whereby the ears and eyes of the host may be focused on one particular point.

25 The host thus forming a single united body, it is impossible either for the brave to advance alone, or for the cowardly to retreat alone. This is the art of handling large masses of men.

26 In night-fighting, then, make much use of signal-fires and drums, and in fighting by day, of flags and banners, as a means of influencing the ears and eyes of your army.

27 A whole army may be robbed of its spirit; a commander-in-chief may be robbed of his presence of mind.

28 Now a soldier's spirit is keenest in the morning; by noonday it has begun to flag; and in the evening, his mind is bent only on returning to camp.

29 A clever general, therefore, avoids an army when its spirit is keen, but attacks it when it is sluggish and inclined to return. This is the art of studying moods.

30 Disciplined and calm, to await the appearance of disorder and hubbub amongst the enemy: – this is the art of retaining self-possession.

31 To be near the goal while the enemy is still far from it, to wait at ease while the enemy is toiling and struggling, to be well-fed while the enemy is famished: – this is the art of husbanding one's strength.

32 To refrain from intercepting an enemy whose banners are in perfect order, to refrain from attacking an army drawn up in calm and confident array: – this is the art of studying circumstances.

33 It is a military axiom not to advance uphill against the enemy, nor to oppose him when he comes downhill.

34 Do not pursue an enemy who simulates flight; do not attack soldiers whose temper is keen.

35 Do not swallow bait offered by the enemy. Do not interfere with an army that is returning home.

36 When you surround an army, leave an outlet free. Do not press a desperate foe too hard.

37 Such is the art of warfare.

CHAPTER EIGHT

VARIATION IN TACTICS

1 Sun Tzu said: In war, the general receives his commands from the sovereign, collects his army and concentrates his forces.

2 When in difficult country, do not encamp. In country where high roads intersect, join hands with your allies. Do not linger in dangerously isolated positions. In hemmed-in situations, you must resort to stratagem. In desperate positions, you must fight.

3 There are roads which must not be followed, armies which must not be attacked, towns which must not be besieged, positions which must not be contested, commands of the sovereign which must not be obeyed.

4 The general who thoroughly understands the advantages that accompany variation of tactics knows how to handle his troops.

5 The general who does not understand these, may be well acquainted with the configuration of the country, yet he will not be able to turn his knowledge to practical account.

6 So, the student of war who is unversed in the art of war of varying his plans, even though he be acquainted with the Five Advantages, will fail to make the best use of his men.

7 Hence in the wise leader's plans, considerations of advantage and of disadvantage will be blended together.

8 If our expectation of advantage be tempered in this way, we may succeed in accomplishing the essential part of our schemes.

9 If, on the other hand, in the midst of difficulties we are always ready to seize an advantage, we may extricate ourselves from misfortune.

10 Reduce the hostile chiefs by inflicting damage on them; and make trouble for them, and keep them constantly engaged; hold out specious allurements, and make them rush to any given point.

11 The art of war teaches us to rely not on the likelihood of the enemy's not coming, but on our own readiness to receive him; not on the chance of his not attacking, but rather on the fact that we have made our position unassailable.

12 There are five dangerous faults which may affect a general:

(1) Recklessness, which leads to destruction;
(2) cowardice, which leads to capture;
(3) a hasty temper, which can be provoked by insults;
(4) a delicacy of honour which is sensitive to shame;
(5) over-solicitude for his men, which exposes him to worry and trouble.

13 These are the five besetting sins of a general, ruinous to the conduct of war.

14 When an army is overthrown and its leader slain, the cause will surely be found among these five dangerous faults. Let them be a subject of meditation.

CHAPTER NINE

THE ARMY ON THE MARCH

1 Sun Tzu said: We come now to the question of encamping the army, and observing signs of the enemy. Pass quickly over mountains, and keep in the neighbourhood of valleys.

2 Camp in high places, facing the sun. Do not climb heights in order to fight. So much for mountain warfare.

3 After crossing a river, you should get far away from it.

4 When an invading force crosses a river in its onward march, do not advance to meet it in mid-stream. It will be best to let half the army get across, and then deliver your attack.

5 If you are anxious to fight, you should not go to meet the invader near a river which he has to cross.

6 Moor your craft higher up than the enemy, and facing the sun. Do not move up-stream to meet the enemy. So much for river warfare.

7 In crossing salt-marshes, your sole concern should be to get over them quickly, without any delay.

8 If forced to fight in a salt-marsh, you should have water and grass near you, and get your back to a clump of trees. So much for operations in salt-marshes.

9 In dry, level country, take up an easily accessible position with rising ground to your right and on your rear, so that the danger may be in front, and safety lie behind. So much for campaigning in flat country.

10 These are the four useful branches of military knowledge which enabled the Yellow Emperor to vanquish four other sovereigns.

11 All armies prefer high ground to low and sunny places to dark.

12 If you are careful of your men, and camp on hard ground, the army will be free from disease of every kind, and this will spell victory.

13 When you come to a hill or a bank, occupy the sunny side, with the slope on your right rear. Thus you will at once act for the benefit of your soldiers and utilize the natural advantages of the ground.

14 When, in consequence of heavy rains up-country, a river which you wish to ford is swollen and flecked with foam, you must wait until it subsides.

15 Country in which there are precipitous cliffs with torrents running between, deep natural hollows, confined places, tangled thickets, quagmires and crevasses, should be left with all possible speed and not approached.

16 While we keep away from such places, we should get the enemy to approach them; while we face them, we should let the enemy have them on his rear.

17 If in the neighbourhood of your camp there should be any hilly country, ponds surrounded by aquatic grass, hollow basins filled with reeds, or woods with thick undergrowth, they must be carefully routed out and searched; for these are places where men in ambush or insidious spies are likely to be lurking.

18 When the enemy is close at hand and remains quiet, he is relying on the natural strength of his position.

19 When he keeps aloof and tries to provoke a battle, he is anxious for the other side to advance.

20 If his place of encampment is easy of access, he is tendering a bait.

21 Movement amongst the trees of a forest shows that the enemy is advancing. The appearance of a number of screens in the midst of thick grass means that the enemy wants to make us suspicious.

22 The rising of birds in their flight is the sign of an ambuscade. Startled beasts indicate that a sudden attack is coming.

23 When there is dust rising in a high column, it is the sign of chariots advancing; when the dust is low, but

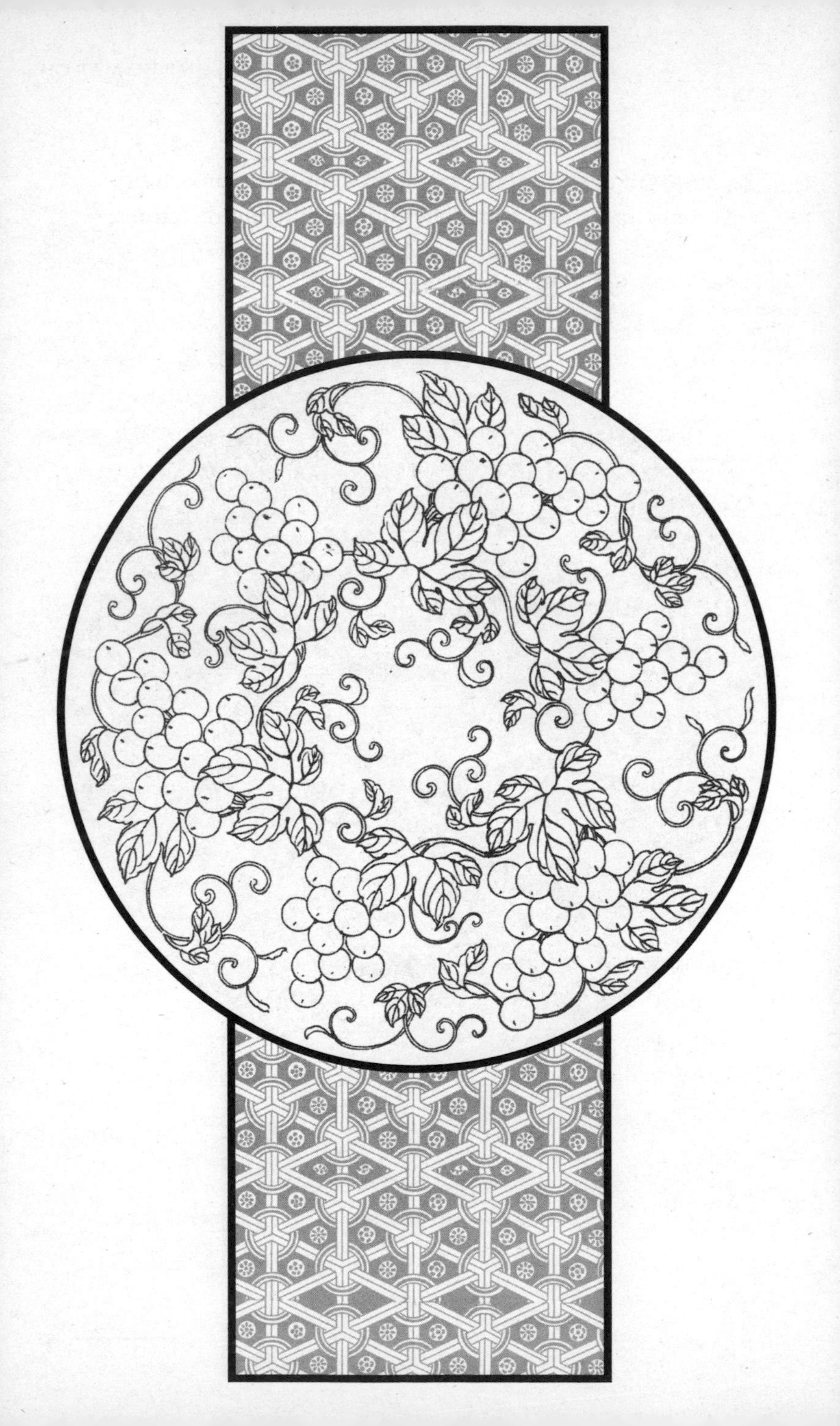

spread over a wide area, it betokens the approach of infantry. When it branches out in different directions, it shows that parties have been sent to collect firewood. A few clouds of dust moving to and fro signify that the army is encamping.

24 Humble words and increased preparations are signs that the enemy is about to advance. Violent language and driving forward as if to the attack are signs that he will retreat.

25 When the light chariots come out first and take up a position on the wings, it is a sign that the enemy is forming for battle.

26 Peace proposals unaccompanied by a sworn covenant indicate a plot.

27 When there is much running about and the soldiers fall into rank, it means that the critical moment has come.

28 When some are seen advancing and some retreating, it is a lure.

29 When the soldiers stand leaning on their spears, they are faint from want of food.

30 If those who are sent to draw water begin by drinking themselves, the army is suffering from thirst.

31 If the enemy sees an advantage to be gained and makes no effort to secure it, the soldiers are exhausted.

32 If birds gather on any spot, it is unoccupied. Clamour by night betokens nervousness.

33 If there is disturbance in the camp, the general's authority is weak. If the banners and flags are shifted about, sedition is afoot. If the officers are angry, it means that the men are weary.

34 When an army feeds its horses with grain and kills its cattle for food, and when the men do not hang their cooking-pots over the camp-fires, showing that they will not return to their tents, you may know that they are determined to fight to the death.

35 The sight of men whispering together in small knots or speaking in subdued tones points to disaffection amongst the rank and file.

36 Too frequent rewards signify that the enemy is at the end of his resources; too many punishments betray a condition of dire distress.

37 To begin by bluster, but afterwards to take fright at the enemy's numbers, shows a supreme lack of intelligence.

38 When envoys are sent with compliments in their mouths, it is a sign that the enemy wishes for a truce.

39 If the enemy's troops march up angrily and remain facing ours for a long time without either joining battle or taking themselves off again, the situation is one that demands great vigilance and circumspection.

40 If our troops are no more in number than the enemy, that is amply sufficient; it only means that no direct attack can be made. What we can do is simply to concentrate all our available strength, keep a close watch on the enemy, and obtain reinforcements.

41 He who exercises no forethought but makes light of his opponents is sure to be captured by them.

42 If soldiers are punished before they have grown attached to you, they will not prove submissive; and, unless submissive, they will be practically useless. If, when the soldiers have become attached to you, punishments are not enforced, they will still be useless.

43 Therefore soldiers must be treated in the first instance with humanity, but kept under control by means of iron discipline. This is a certain road to victory.

44 If in training soldiers commands are habitually enforced, the army will be well-disciplined; if not, its discipline will be bad.

45 If a general shows confidence in his men but always insists on his orders being obeyed, the gain will be mutual.

CHAPTER TEN

TERRAIN

1 Sun Tzu said: We may distinguish six kinds of terrain, to wit:

(1) Accessible ground;
(2) entangling ground;
(3) temporizing ground;
(4) narrow passes;
(5) precipitous heights;
(6) positions at a great distance from the enemy.

2 Ground which can be freely traversed by both sides is called accessible.

3 With regard to ground of this nature, be before the enemy in occupying the raised and sunny spots, and carefully guard your line of supplies. Then you will be able to fight with advantage.

4 Ground which can be abandoned but is hard to re-occupy is called entangling.

5 From a position of this sort, if the enemy is unprepared, you may sally forth and defeat him. But if the enemy is prepared for your coming, and you fail to defeat him, then, return being impossible, disaster will ensue.

6 When the position is such that neither side will gain by making the first move, it is called temporizing ground.

7 In a position of this sort, even though the enemy should offer us an attractive bait, it will be advisable not to stir forth, but rather to retreat, thus enticing the enemy in his turn; then, when part of his army has come out, we may deliver our attack with advantage.

8 With regard to narrow passes, if you can occupy them first, let them be strongly garrisoned and await the advent of the enemy.

9 Should the army forestall you in occupying a pass, do not go after him if the pass is fully garrisoned, but only if it is weakly garrisoned.

10 With regard to precipitous heights, if you are beforehand with your adversary, you should occupy the raised and sunny spots, and there wait for him to come up.

11 If the enemy has occupied them before you, do not follow him, but retreat and try to entice him away.

12 If you are situated at a great distance from the enemy, and the strength of the two armies is equal, it is not easy to provoke a battle, and fighting will be to your disadvantage.

13 These six are the principles connected with Earth. The general who has attained a responsible post must be careful to study them.

14 Now an army is exposed to six calamities, not arising from natural causes, but from faults for which the general is responsible. These are:

(1) Flight;
(2) insubordination;
(3) collapse;
(4) ruin;
(5) disorganization;
(6) rout.

15 Other conditions being equal, if one force is hurled against another ten times its size, the result will be the flight of the former.

16 When the common soldiers are too strong and their officers too weak, the result is insubordination. When the officers are too strong and the common soldiers too weak, the result is collapse.

17 When the higher officers are angry and insubordinate, and on meeting the enemy give battle on their own account from a feeling of resentment, before the commander-in-chief can tell whether or not he is in a position to fight, the result is ruin.

18 When the general is weak and without authority; when his orders are not clear and distinct; when there are no fixed duties assigned to officers and men, and the ranks are formed in a slovenly haphazard manner, the result is utter disorganization.

19 When a general, unable to estimate the enemy's strength, allows an inferior force to engage a larger one, or hurls a weak detachment against a powerful one, and neglects to place picked soldiers in the front rank, the result must be a rout.

20 These are six ways of courting defeat, which must be carefully noted by the general who has attained a responsible post.

21 The natural formation of the country is the soldier's best ally; but a power of estimating the adversary, of controlling the forces of victory, and of shrewdly calculating difficulties, dangers and distances, constitutes the test of a great general.

22 He who knows these things, and in fighting puts his knowledge into practice, will win his battles. He who knows them not, nor practises them, will surely be defeated.

23 If fighting is sure to result in victory, then you must fight, even though the ruler forbid it; if fighting will not result in victory, then you must not fight even at the ruler's bidding.

24 The general who advances without coveting fame and retreats without fearing disgrace, whose only thought is to protect his country and do good service for his sovereign, is the jewel of the kingdom.

25 Regard your soldiers as your children, and they will follow you into the deepest valleys; look upon them as your own beloved sons, and they will stand by you even unto death.

26 If, however, you are indulgent, but unable to make your authority felt; kind-hearted, but unable to enforce your commands; and incapable, moreover, of quelling disorder: then your soldiers must be likened to spoilt children; they are useless for any practical purpose.

27 If we know that our own men are in a condition to attack, but are unaware that the enemy is not open to attack, we have gone only halfway towards victory.

28 If we know that the enemy is open to attack, but are unaware that our own men are not in a condition to attack, we have gone only halfway towards victory.

29 If we know that the enemy is open to attack, and also know our men are in a condition to attack, but are unaware that the nature of the ground makes fighting impracticable, we have still gone only halfway towards victory.

30 Hence the experienced soldier, once in motion, is never bewildered; once he has broken camp, he is never at a loss.

31 Hence the saying: If you know the enemy and know yourself, your victory will not stand in doubt; if you know Heaven and know Earth, you may make your victory complete.

CHAPTER ELEVEN

THE NINE SITUATIONS

1 Sun Tzu said: The art of war recognizes nine varieties of ground:

(1) Dispersive ground;
(2) facile ground;
(3) contentious ground;
(4) open ground;
(5) ground of intersecting highways;
(6) serious ground;
(7) difficult ground;
(8) hemmed-in ground;
(9) desperate ground.

2 When a chieftain is fighting in his own territory, it is dispersive ground.

3 When he has penetrated into hostile territory, but to no great distance, it is facile ground.

4 Ground the possession of which imports great advantage to either side, is contentious ground.

5 Ground on which each side has liberty of movement is open ground.

6 Ground which forms the key to three contiguous states, so that he who occupies it first has most of the Empire at his command, is a ground of intersecting highways.

7 When an army has penetrated into the heart of a

hostile country, leaving a number of fortified cities in its rear, it is serious ground.

8 Mountain forests, rugged steeps, marshes and fens – all country that is hard to traverse: this is difficult ground.

9 Ground which is reached through narrow gorges, and from which we can only retire by tortuous paths, so that a small number of the enemy would suffice to crush a large body of our men: this is hemmed-in ground.

10 Ground on which we can only be saved from destruction by fighting without delay, is desperate ground.

11 On dispersive ground, therefore, fight not. On facile ground, halt not. On contentious ground, attack not.

12 On open ground, do not try to block the enemy's way. On the ground of intersecting highways, join hands with your allies.

13 On serious ground, gather in plunder. In difficult ground, keep steadily on the march.

14 On hemmed-in ground, resort to stratagem. On desperate ground, fight.

15 Those who were called skilful leaders of old knew how to drive a wedge between the enemy's front

and rear; to prevent co-operation between his large and small divisions; to hinder the good troops from rescuing the bad, the officers from rallying their men.

16 When the enemy's men were united, they managed to keep them in disorder.

17 When it was to their advantage, they made a forward move; when otherwise, they stopped still.

18 If asked how to cope with a great host of the enemy in orderly array and on the point of marching to the attack, I should say: 'Begin by seizing something which your opponent holds dear; then he will be amenable to your will.'

19 Rapidity is the essence of war: take advantage of the enemy's unreadiness, make your way by unexpected routes, and attack unguarded spots.

20 The following are the principles to be observed by an invading force: The further you penetrate into a country, the greater will be the solidarity of your troops, and thus the defenders will not prevail against you.

21 Make forays in fertile country in order to supply your army with food.

22 Carefully study the well-being of your men, and do not overtax them. Concentrate your energy and hoard your strength. Keep your army continually on the move, and devise unfathomable plans.

23 Throw your soldiers into positions whence there is no escape, and they will prefer death to flight. If they will face death, there is nothing they may not achieve. Officers and men alike will put forth their uttermost strength.

24 Soldiers when in desperate straits lose the sense of fear. If there is no place of refuge, they will stand firm. If they are in hostile country, they will show a stubborn front. If there is no help for it, they will fight hard.

25 Thus, without waiting to be marshalled, the soldiers will be constantly on the *qui vive*; without waiting to be asked, they will do your will; without restrictions, they will be faithful; without giving orders, they can be trusted.

26 Prohibit the taking of omens, and do away with superstitious doubts. Then, until death itself comes, no calamity need be feared.

27 If our soldiers are not overburdened with money, it is not because they have a distaste for riches; if their lives are not unduly long, it is not because they are disinclined to longevity.

28 On the day they are ordered out to battle, your soldiers may weep, those sitting up bedewing their garments, and those lying down letting the tears run down their cheeks. But let them once be brought to bay, and they will display the courage of a Chu or a Kuei.

29 The skilful tactician may be likened to the shuai-jan. Now the shuai-jan is a snake that is found in the Ch'ang mountains. Strike at its head, and you will be attacked by its tail; strike at its tail, and you will be attacked by its head; strike at its middle, and you will be attacked by head and tail both.

30 Asked if an army can be made to imitate the shuai-jan, I should answer, 'Yes'. For the men of Wu and the men of Yueh are enemies; yet if they are crossing a river in the same boat and are caught by a storm, they will come to each other's assistance just as the left hand helps the right.

31 Hence it is not enough to put one's trust in the tethering of horses, and the burying of chariot wheels in the ground.

32 The principle on which to manage an army is to set up one standard of courage which all must reach.

33 How to make the best of both strong and weak, that is a question involving the proper use of ground.

34 Thus the skilful general conducts his army just as though he were leading a single man, willy-nilly, by the hand.

35 It is the business of a general to be quiet and thus ensure secrecy; upright and just, and thus maintain order.

36 He must be able to mystify his officers and men by false reports and appearances, and thus keep them in total ignorance.

37 By altering his arrangements and changing his plans, he keeps the enemy without definite knowledge. By shifting his camp and taking circuitous routes, he prevents the enemy from anticipating his purpose.

38 At the critical moment, the leader of an army acts like one who has climbed up a height and then kicks away the ladder behind him. He carries his men deep into hostile territory before he shows his hand.

39 He burns his boats and breaks his cooking-pots; like a shepherd driving a flock of sheep, he drives his men this way and that, and nothing knows whither he is going.

40 To muster his host and bring it into danger: – this may be termed the business of the general.

41 The different measures suited to the nine varieties of ground; the expediency of aggressive or defensive tactics; and the fundamental laws of human nature: these are things that must most certainly be studied.

42 When invading hostile territory, the general principle is that penetrating deeply brings cohesion; penetrating but a short way means dispersion.

43 When you leave your own country behind, and take your army across neighbourhood territory, you find yourself on critical ground. When there are means of communication on all four sides, the ground is one of intersecting highways.

44 When you penetrate deeply into a country, it is serious ground. When you penetrate but a little way, it is facile ground.

45 When you have the enemy's strongholds in your rear, and narrow passes in front, it is hemmed-in ground. When there is no place of refuge at all, it is desperate ground.

46 Therefore, on dispersive ground, I would inspire my men with unity of purpose. On facile ground, I would see that there is close connection between all parts of my army.

47 On contentious ground, I would hurry up my rear.

48 On open ground, I would keep a vigilant eye on my defences. On ground of intersecting highways, I would consolidate my alliances.

49 On serious ground, I would try to ensure a continuous stream of supplies. On difficult ground, I would keep pushing on along the road.

50 On hemmed-in ground, I would block any way of retreat. On desperate ground, I would proclaim to my soldiers the hopelessness of saving their lives.

51 For it is the soldier's disposition to offer an obstinate resistance when surrounded, to fight hard when he cannot help himself, and to obey promptly when he has fallen into danger.

52 We cannot enter into alliance with neighbouring princes until we are acquainted with their designs. We are not fit to lead an army on the march unless we are familiar with the face of the country – its mountains and forests, its pitfalls and precipices, its marshes and swamps. We shall be unable to turn natural advantages to account unless we make use of local guides.

53 To be ignorant of any one of the following four or five principles does not befit a warlike prince.

54 When a warlike prince attacks a powerful state, his generalship shows itself in preventing the concentration of the enemy's forces. He overawes his opponents, and their allies are prevented from joining against him.

55 Hence he does not strive to ally himself with all and sundry, nor does he foster the power of other states. He carries out his own secret designs, keeping his antagonists in awe. Thus he is able to capture their cities and overthrow their kingdoms.

56 Bestow rewards without regard to rule, issue orders without regard to previous arrangements; and you will be able to handle a whole army as though you had to do with but a single man.

57 Confront your soldiers with the deed itself; never let them know your design. When the outlook is bright, bring it before their eyes; but tell them nothing when the situation is gloomy.

58 Place your army in deadly peril, and it will survive; plunge it into desperate straits, and it will come off in safety.

59 For it is precisely when a force has fallen into harm's way that it is capable of striking a blow for victory.

60 Success in warfare is gained by carefully accommodating ourselves to the enemy's purpose.

61 By persistently hanging on the enemy's flank, we shall succeed in the long run in killing the commander-in-chief.

62 This is called the ability to accomplish a thing by sheer cunning.

63 On the day that you take up your command, block the frontier passes, destroy the official tallies, and stop the passage of all emissaries.

64 Be stern in the council-chamber, so that you may control the situation.

65 If the enemy leaves a door open, you must rush in.

66 Forestall your opponent by seizing what he holds dear, and subtly contrive to time his arrival on the ground.

67 Walk in the path defined by rule, and accommodate yourself to the enemy until you can fight a decisive battle.

68 At first, then, exhibit the coyness of a maiden, until the enemy gives you an opening; afterwards emulate the rapidity of a running hare, and it will be too late for the enemy to oppose you.

CHAPTER TWELVE

ATTACK BY FIRE

1 Sun Tzu said: There are five ways of attacking with fire. The first is to burn soldiers in their camp; the second is to burn stores; the third is to burn baggage trains; the fourth is to burn arsenals and magazines; the fifth is to hurl dropping fire amongst the enemy.

2 In order to carry out an attack, we must have means available. The material for raising fire should always be kept in readiness.

3 There is a proper season for making attacks with fire, and special days for starting a conflagration.

4 The proper season is when the weather is very dry; the special days are those when the moon is in the constellations of the Sieve, the Wall, the Wing or the Cross-bar; for these four are all days of rising wind.

5 In attacking with fire, one should be prepared to meet five possible developments:

6 (1) When fire breaks out inside the enemy's camp, respond at once with an attack from without.

7 (2) If there is an outbreak of fire, but the enemy's soldiers remain quiet, bide your time and do not attack.

8 (3) When the force of the flames has reached its height, follow it up with an attack, if that is practicable; if not, stay where you are.

9 (4) If it is possible to make an assault with fire from without, do not wait for it to break out within, but deliver your attack at a favourable moment.

10 (5) When you start a fire, be to windward of it. Do not attack from the leeward.

11 A wind that rises in the daytime lasts long, but a night breeze soon falls.

12 In every army, the five developments connected with fire must be known, the movements of the stars calculated, and a watch kept for the proper days.

13 Hence those who use fire as an aid to the attack show intelligence; those who use water as an aid to the attack gain an accession of strength.

14 By means of water, an enemy may be intercepted, but

not robbed of all his belongings.

15 Unhappy is the fate of one who tries to win his battles and succeed in his attacks without cultivating the spirit of enterprise; for the result is waste of time and general stagnation.

16 Hence the saying: The enlightened ruler lays his plans well ahead; the good general cultivates his resources.

17 Move not unless you see an advantage; use not your troops unless there is something to be gained; fight not unless the position is critical.

18 No ruler should put troops into the field merely to gratify his own spleen; no general should fight a battle simply out of pique.

19 If it is to your advantage, make a forward move; if not, stay where you are.

20 Anger may in time change to gladness; vexation may be succeeded by content.

21 But a kingdom that has once been destroyed can never come again into being; nor can the dead ever be brought back to life.

22 Hence the enlightened ruler is heedful, and the good general full of caution. This is the way to keep a country at peace and an army intact.

CHAPTER THIRTEEN

THE USE OF SPIES

1 Sun Tzu said: Raising a host of a hundred thousand men and marching them great distances entails heavy loss on the people and a drain on the resources of the State. The daily expenditure will amount to a thousand ounces of silver. There will be commotion at home and abroad, and men will drop down exhausted on the highways. As many as seven hundred thousand families will be impeded in their labour.

2 Hostile armies may face each other for years, striving for the victory which is decided in a single day. This being so, to remain in ignorance of the enemy's condition simply because one grudges the outlay of a hundred ounces of silver in honours and emoluments, is the height of inhumanity.

3 One who acts thus is no leader of men, no present help to his sovereign, no master of victory.

4 Thus, what enables the wise sovereign and the good general to strike and conquer, and achieve things beyond the reach of ordinary men, is foreknowledge.

5 Now this foreknowledge cannot be elicited from spirits; it cannot be obtained inductively from experience, nor by any deductive calculation.

6 Knowledge of the enemy's dispositions can only be obtained from other men.

7 Hence the use of spies, of whom there are five classes:
(1) Local spies;
(2) inward spies;
(3) converted spies;
(4) doomed spies;
(5) surviving spies.

8 When these five kinds of spy are all at work, none can discover the secret system. This is called 'divine manipulation of the threads'. It is the sovereign's most precious faculty.

9 Having local spies means employing the services of the inhabitants of a district.

10 Having inward spies, means making use of officials of the enemy.

11 Having converted spies, means getting hold of the enemy's spies and using them for our own purposes.

12 Having doomed spies, doing certain things openly for purposes of deception, and allowing our spies to know of them and report them to the enemy.

13 Surviving spies, finally, are those who bring back news from the enemy's camp.

14 Hence it is that with none in the whole army are more intimate relations to be maintained than with spies. None should be more liberally rewarded. In no

other business should greater secrecy be preserved.

15 Spies cannot be usefully employed without a certain intuitive sagacity.

16 They cannot be properly managed without benevolence and straightforwardness.

17 Without subtle ingenuity of mind, one cannot make certain of the truth of their reports.

18 Be subtle! be subtle! and use your spies for every kind of business.

19 If a secret piece of news is divulged by a spy before the time is ripe, he must be put to death together with the man to whom the secret was told.

20 Whether the object be to crush an army, to storm a city, or to assassinate an individual, it is always necessary to begin by finding out the names of the attendants, the aides-de-camp, and door-keepers and sentries of the general in command. Our spies must be commissioned to ascertain these.

21 The enemy's spies who have come to spy on us must be sought out, tempted with bribes, led away and comfortably housed. Thus they will become converted spies and available for our service.

龍

22 It is through the information brought by the converted spy that we are able to acquire and employ local and inward spies.

23 It is owing to his information, again, that we can cause the doomed spy to carry false tidings to the enemy.

24 Lastly, it is by his information that the surviving spy can be used on appointed occasions.

25 The end and aim of spying in all its five varieties is knowledge of the enemy; and this knowledge can only be derived, in the first instance, from the converted spy. Hence it is essential that the converted spy be treated with the utmost liberality.

26 Of old, the rise of the Yin dynasty was due to I Chih who had served under the Hsia. Likewise, the rise of the Chou dynasty was due to Lu Ya who had served under the Yin.

27 Hence it is only the enlightened ruler and the wise general who will use the highest intelligence of the army for purposes of spying and thereby they achieve great results. Spies are a most important element in warfare, because on them depends an army's ability to move.